Military Technology

Ron Fridell

LERNER PUBLICATIONS COMPANY
MINNEAPOLIS

Lerner Publications Company
A division of Lerner Publishing Group, Inc.
241 First Avenue North
Minneapolis, MN 55401 U.S.A.

Website address: www.lernerbooks.com

Library of Congress Cataloging-in-Publication Data

Fridell, Ron.
Military technology / by Ron Fridell.
p. cm. — (Cool science)
Includes bibliographical references and index.
ISBN-13: 978–0–8225–6769–1 (lib. bdg. : alk. paper)
ISBN-10: 0–8225–6769–5 (lib. bdg. : alk. paper)
1. Weapons system—History—Juvenile literature 2. Military art and science—Effect of technological innovations on—Juvenile literature. I. Title.
UF500.F739 2008
623—dc22 2006019404

Manufactured in the United States of America
1 2 3 4 5 6 – JR – 13 12 11 10 09 08

Table of Contents

The year is A.D. 378. The place is Adrianople, Turkey. Two ancient armies, the Goths and the Byzantines, prepare to do battle.

Each army has thousands of warriors. Most are infantry—foot soldiers. They carry swords and shields. The rest are on horseback with spears, bows, and arrows.

The Goth horsemen attack first, surrounding the Byzantine infantry. They ride in ever-tightening circles around their enemy. Soon the Byzantine warriors are crowded together too tightly to even raise their swords or shields. Clouds of dust from the horses' hooves choke and blind the Byzantine infantry. The Goth archers send showers of arrows down upon their enemy. Then the infantry joins the attack. The Goth army wins.

The Goths' victory came from their superior use of military technology, the tools of warfare. A military is a collection of people, machines, and equipment that form an army. An army uses military technology to protect its own soldiers and defeat its enemy. After squeezing the enemy with their horses, the Goths used their weapons to finish them off.

Military technology has changed a lot since the battle at Adrianople. Modern warriors travel by battleship, helicopter, airplane, and tank. Instead of carrying shields, they wear flexible body armor. And modern warriors don't use spears and arrows. They use automatic rifles, laser-guided missiles, and other scientifically advanced weapons to fight and defeat their enemies.

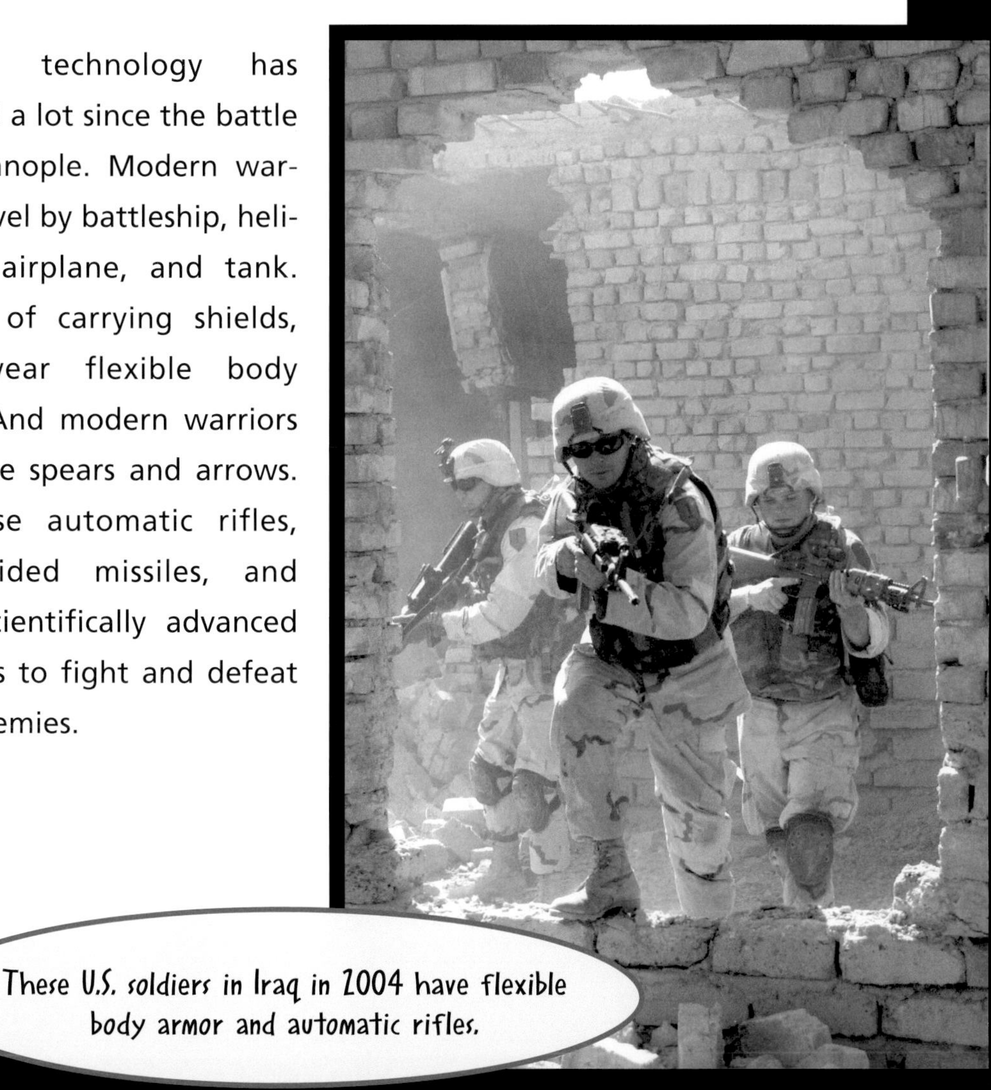

These U.S. soldiers in Iraq in 2004 have flexible body armor and automatic rifles.

Bullets and Heat Beams

For thousands of years, soldiers used their muscles to power weapons such as swords, spears, bows and arrows, and crossbows. Then, in the 1200s, came gunpowder. This mixture of chemicals could be hit or burned to make it explode. Gunpowder changed warfare forever.

Using Gunpowder

The front-loading cannon was one of the first weapons to use gunpowder. The powder was

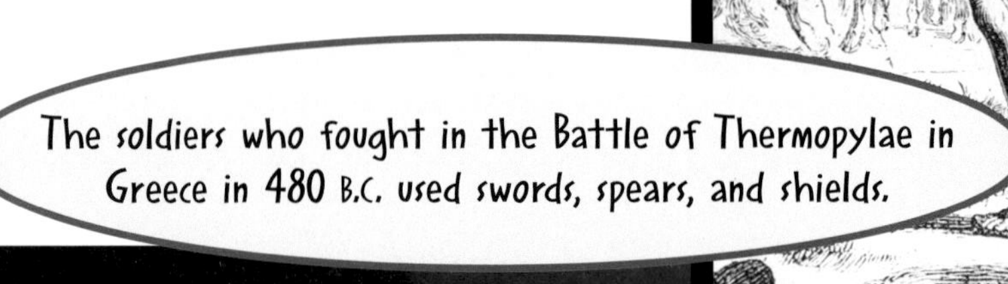

The soldiers who fought in the Battle of Thermopylae in Greece in 480 B.C. used swords, spears, and shields.

Both armies used cannons in the American Civil War (1861–1865). Here Union (Northern) officers get ready to fire a cannon.

poured down the barrel and packed tightly. Then a large, heavy cannonball was dropped in.

IT'S A FACT!

The first cannonballs were made of stone and weighed about 450 pounds (204 kilograms).

A fuse ran in through a hole in the back of the barrel to where the gunpowder was packed. Soldiers lit the fuse, creating fire to ignite the gunpowder. The power of the explosion launched the cannonball out of the barrel with a burst of noise and flame.

Before long, people developed handheld guns. These muskets were loaded like cannons, from the front. Their barrels were smooth inside. When fired, the cannonball or bullet would bang around inside before shooting out. They were not very accurate, but they could still be deadly at close range.

Around 1665 the first accurate rifle was invented. Rifles had a spiral groove running all the way along the inside of the barrel. The groove guided the bullet along. The bullet came out spinning, and it flew straight and far. Some Americans used rifles in the American Revolution (1775–1783). In the American Civil War, most soldiers used rifles.

Rifles were also used in the Civil War, such as in the Battle of Kennesaw Mountain in Georgia in 1864.

Wooden Victims

The deadliest battle in the American Civil War was the Battle of Gettysburg (July 1–3, 1863). More than 50,000 soldiers were killed, wounded, or captured. Soldiers were not the only victims of the cannon shells and rifle bullets. Nearby trees were killed from the shells and lead bullets that tore through them. Years later, still more of the battlefield trees were dying. The cause was said to be lead poisoning from the bullets in their trunks.

At first, rifles shot round bullets and were loaded from the muzzle, just like muskets. Then came bullet cartridges. These neat packages held both a lead bullet and gunpowder packed tight inside a metal shell. The shells were easy to load. The soldier inserted them from the breech, or rear, of the weapon.

Automatic Weapons

The first muskets and rifles shot one bullet at a time. After each shot, the shooter had to stop to reload. Firing a single-shot weapon took time, skill, and patience. To hit his target, the shooter had to be well trained. Then came firearms that could be loaded with more than one bullet. But they still shot one at a time.

By the late 1800s, soldiers were using automatic rifles and machine guns. These guns could fire many bullets with a single pull of the trigger. A machine gunner's weapon fired hundreds of bullets each minute. He could point the weapon in the general direction of his enemy and fire. Even poorly-trained shooters could hit their targets. Automatic weapons made war a far more deadly business.

For modern soldiers, the most common military weapon is the AK-47 assault rifle. The AK-47 is not a very accurate weapon. It shakes and rattles when fired. But it can unleash about 700 shots per minute. As many as 100 million of these inexpensive rifles are in use in conflicts around the world. That's about one AK-47 for every six people on Earth.

The AK-47 is very reliable and is easy to buy. For this reason, it is the weapon of choice for most terrorists. Osama bin Laden, head of the al-Qaeda terrorist group, has called the AK-47 the terrorist's most important weapon.

Reliability

Why do so many warriors use the AK-47? Reliability. Above all else, a combat rifle must be reliable. A soldier's life depends upon it. In his book about the AK-47, author Larry Kahaner writes, "[Y]ou can drag it through mud, leave it buried in the sand and take it out a year later, kick it with your boot, and it will fire like it was cleaned that morning."

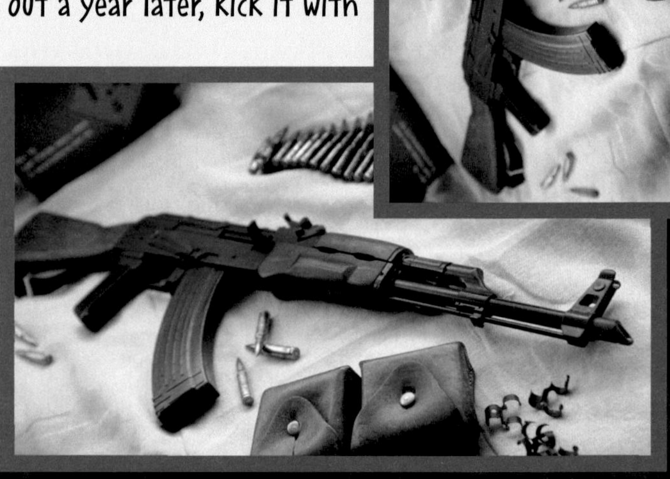

The AK-47 is shown with ammunition.

The Corner Shot

There's an old saying: "You can't hit what you can't see." But that's not totally true anymore. With a new special-purpose gun, the Corner Shot, a shooter can actually shoot around corners while remaining behind cover.

The Corner Shot was specially designed for conflicts involving hostages and terrorists. It looks like a rifle with a hinge in the middle. The front end holds a pistol, a digital camera, and a flashlight. On the back end are controls for the camera and light, a camera screen, and a trigger.

The Corner Shot is hinged in the middle so that the shooter can bend the front part sideways. Then the pistol and camera can be pointed around the corner. By watching the camera screen, the hidden shooter

Amos Golan (*left*), an Israeli inventor, developed the Corner Shot.

can see and fire at a target around that corner. With the hinge snapped back in place, the gun can be used as a conventional rifle.

Fire and Heat

Some weapons shoot things besides bullets. A flamethrower projects a long stream of fire. The weapon consists of a backpack and gun. The backpack has two or three tanks. One holds a compressed gas, usually nitrogen. The other tanks contain a flammable liquid. The tanks are all connected. The gas propels (moves) the liquid through a flexible pipe hooked up to the gun.

When the trigger is pulled, the liquid is ignited and shot out the nozzle. Instead of actual flames, the weapon shoots a stream of burning

A German solider uses a flame thrower weapon during military training in 1941.

liquid. Soldiers used flamethrowers in World War I (1914–1918) and in World War II (1939–1945). They were especially effective in clearing out soldiers from hard-to-reach caves and bunkers.

A new weapon is being tested. It fires invisible heat rays instead of bullets or fire. The Active Denial System (ADS) is a directed energy weapon. It sends out highly concentrated energy in a directed stream. The ADS is mounted on a vehicle, such as a Jeep or Humvee. A large dish antenna directs an invisible ray of intense energy in a straight line toward the target.

The ADS will be used for crowd control. The goal is to cause pain without injury. It will help U.S. soldiers break up angry or suspicious groups of people without actually hurting anyone. The ray delivers a harmless but painful punch over a long distance. People several hundred yards (meters) away feel the same amount of pain as people standing only a few feet away. It's like they've been hit by a blast from a hot oven. They must duck and run for cover.

Zapped

In January 2007, the Department of Defense tested its new ADS heat ray weapon on volunteers at a Georgia air force base. Two dozen soldiers and a newspaper reporter were hit.

The moment the ray hit them, the soldiers scattered in all directions. The reporter said it felt as if his clothes were on fire, but he wasn't actually burned. No one was. The invisible ray just barely penetrates the surface of the skin. It zapps nerve endings with a fierce jolt of pain, but not for long. The pain stops the moment the person runs out of range of the ray.

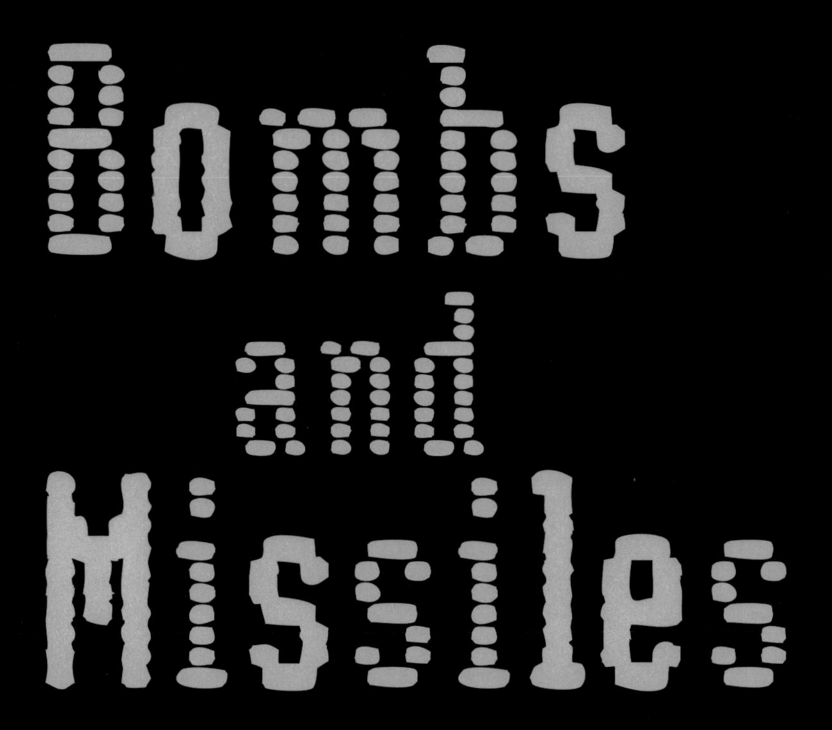

Bombs and Missiles

Rifles, flamethrowers, and heat rays send bullets, liquid fire, and energy streams speeding toward a target. Bombs and missiles are different. Their explosive energy flies out in all directions at once. The result is a shock wave that smashes anything and everything in its path. Bombs are used for many purposes, such as killing the enemy, destroying their hideouts, and destroying their ability to make war.

Little and Big

Bombs come in all shapes and sizes. Hand grenades are small bombs with metal shells. They are called antipersonnel devices—they are designed to kill people. Modern hand grenades were first used in World War I. A fuse starts burning as the soldier hurls the grenade. A few

This soldier throws a hand grenade during World War II.

seconds later, the fuse fires the explosives inside. Deadly shrapnel—pieces of the metal shell—fly outward. Grenades can also be fired from rifles, guns, or special grenade launchers.

A modern version of a small bomb is the EFP, short for "explosively formed projectile." Warriors on both sides of the Iraq war (which started in 2003) have used EFPs. The U.S. version is small enough to fit in a pocket. It weighs about 2 pounds (0.9 kg). One side is flat. The other is concave (curved inward). When the EFP explodes, the concave end blows outward and melts. This turns it into a bullet-shaped fragment that can slice through almost anything.

The EFP has been nicknamed the Tank Killer. It can be buried in the middle of a road or alongside it. EFPs come equipped with a magnetic sensor. When it senses a vehicle passing by, the sensor triggers the bomb to explode. It can penetrate the thick steel armor on a tank from 25 feet (7.6 meters) away.

Compare the 2-pound EFP (0.9 kg) with the 21,500-pound (9,752 kg) Massive Ordnance Air Blast bomb (MOAB). While a soldier can carry an EFP in his pocket, a MOAB (nicknamed the Mother of All Bombs) must be carried and dropped from a U.S. C-130 Hercules transport plane. The C-130 is about the size of a passenger airliner. The MOAB was built for use against large formations of troops or against underground bunkers. It has never been used in combat.

A MOAB bomb at an air force base in Florida

Smart Bombs

It is not easy to get a free-falling bomb dropped from a plane thousands of feet (meters) in the air to hit its target. The U.S. military helped solve this problem with precision-guided smart bombs.

A key to the success of modern smart bombs is the Global Positioning System (GPS). This network of more than two dozen satellites orbits Earth. The satellites send out radio signals. People with GPS receivers can pick up these signals. The signals give the exact location of the receiver. People can look at a GPS receiver to see exactly where on Earth they are.

When a GPS receiver is added to a bomb, the bomb always "knows" its position on Earth. Then a computer is added—a computer that knows exactly where the target is located. Together, the GPS and the computer can guide the smart bomb to that exact spot.

IT'S A FACT!

How accurate are smart bombs? When dropped from planes thousands of feet in the air, the bombs can be guided to hit within 10 feet (3 meters) of their target.

Another key to smart-bomb success is laser light. A laser-guided bomb (LGB) uses a beam of laser light to tag, or mark, its target. The light reflecting back from the target is known as sparkle. The head of the weapon senses the sparkle and signals its control mechanism to guide it to that spot. LGBs are 100 times more accurate than unguided, free-falling, "dumb" bombs.

Smart Missiles

A missile is a rocket that carries an explosive warhead. Missiles are equipped with some of the same guidance systems as smart bombs. Some missiles are used to destroy enemy aircraft, tanks, and ships. Larger missiles perform the same job as bombs dropped from aircraft.

The AIM-9 Sidewinder was one of the first guided missiles. Sidewinders have been around since 1953. Many air forces still use them. It was named after the Sidewinder snake for two reasons. First, early versions had an odd, twisting, snakelike flight pattern. Second, Sidewinder missiles detect their targets the same way that Sidewinder snakes detect their prey, by the heat the target gives off. It's a heat-seeking missile. Its guidance system homes in on infrared (IR) heat waves given off by the target—for example, an aircraft's engine. Sidewinders are launched from aircraft and can travel about 11 miles (18 kilometers) under their own power.

An AGM-65 Maverick missile is launched from an F-4E Phantom II aircraft.

The AGM-65 Maverick is another guided missile. Like the Sidewinder, the Maverick is launched from an aircraft and is powered by its own rocket motor. The Maverick can use three different guidance systems. Some use an electro-optical television system. After firing the missile, the pilot views the target through a camera mounted in the missile's head. Then the pilot locks the missile in on the target. The Maverick also comes in laser-guided and heat-seeking models.

Early missiles such as the Sidewinder could fly only a few miles. Then came long-range cruise missiles such as the Tomahawk. The U.S. military first carried this missile in 1984. It can carry a large warhead 1,000 miles (1,609 km) under its own power. It comes equipped with a jet engine and a GPS to bring it to its target. The Tomahawk can be launched from a submarine, a warship, or from land.

Shock and Awe

When U.S.-led forces attacked Iraq in 2003, they used a military strategy called shock and awe. On March 21, U.S. forces unleashed a seven-minute barrage of precision-guided bombs and missiles on Iraq's capital city of Baghdad. The attack continued. Within the first 24 hours, about 1,500 bombs and missiles hit the city.

Journalist May Ying Welsh was there. She called the experience "terrifying" and "heart-stopping." As a missile approaches, she said, "It shrieks by you and you have no idea where it's going to fall."

The goal was to use overwhelming force to terrify the enemy into immediately surrendering. This did not happen. But Baghdad did eventually fall to U.S.-led forces in April. Iraqi dictator Saddam Hussein's army was defeated. The United States remains in Iraq at war with insurgents, or rebels, who oppose the new government there.

A Tomahawk missile is launched toward Baghdad, Iraq, from the USS Porter *destroyer on March 22, 2003.*

The Deadliest of All

The bombs and missiles in this chapter can bring terrible destruction. But none can compare with a nuclear bomb.

With this deadliest of all explosive devices, millions of powerful explosions happen all at once. These explosions are atoms splitting apart. Nuclear bombs are the most destructive military technology on Earth. When a nuclear bomb explodes, it destroys everything within several miles.

Nuclear bombs have been used only in World War II. The United States was determined to bring this long and terrible war to a quick end.

In 1945, U.S. aircraft dropped nuclear bombs on two Japanese cities: Hiroshima, on August 6, and Nagasaki, on August 9. These were big cities, but in seconds, they were leveled. Within seconds hundreds of thousands of Japanese soldiers and civilians were dead.

On August 15, 1945, Japan surrendered. The war was over. Many lives were lost. Many others were saved. But a new age had begun—the nuclear age.

The United States dropped nuclear bombs on the Japanese cities of Hiroshima and Nagasaki (*above*) in 1945.

Soon after the war ended, the United States' greatest enemy, the Soviet Union, had developed its own atomic bomb. The two nations became involved in a standoff known as the Cold War (1945–1991). They never actually fought, but tensions remained high between them.

Over the decades, the United States and Soviet Union built more powerful nuclear weapons. By the 1960s, both countries had the power to totally destroy each other—and the rest of the world with them. The Cold War ended with the collapse of the Soviet Union in 1991. But the threat of a nuclear disaster remains. Many people worry that a terrorist organization such as al-Qaeda might get its hands on a nuclear bomb.

Dragon Skin and Death Rays

Attack—wounding and killing the enemy—is one half of war. Defense is the other half. For a soldier in combat, defensive technologies can mean the difference between life and death.

Shields and Helmets

Think of how important the heart and brain are. A sharp blow to either one could cause serious, even deadly injury. That's why ancient defensive equipment focused on protecting the chest and head.

This ivory carving shows a group of Roman soldiers with their helmets and shields in 200 B.C.

Ancient Ruins

You might ask, *How do we know so much about ancient weapons and shields?* Archaeologists have dug them up from battlefields and from the ruins of ancient cities. In China, for example, archaeologists have uncovered the site of the 2,200-year-old city of Chang'an. About 240,000 people once lived there. Discoveries included a weapons warehouse with iron swords, bows and arrows, and armor.

Infantry soldiers were especially at risk. So they would often fight with a weapon in one hand and a shield in the other. Ancient Roman soldiers used shields made of three layers of plywood joined by glue made from animal hides. The outer edges were covered in bronze strips or rawhide.

Most early armor was made of animal hides. Later materials for armor included leather, bone, bronze, and eventually steel. Each new material gave the warrior better protection.

Eleventh-century knights in their armor

When we think of armor, we usually picture knights. These European warriors fought battles during the Middle Ages (A.D. 500–1450). Knights often wore chain mail with shiny plate armor over it.

Chain mail is made of small metal rings linked together.

This kind of armor is flexible. But sometimes cuts and blows from swords and axes got through.

Plate armor is made of solid iron plates. This armor is not flexible. It's hard to move around in. Yet plate armor protected the knight better than mail.

IT'S A FACT!

The earliest known armor from China was made of turtle shells held together with rope.

Deadly Gaps

No piece of defensive equipment is perfect. Even solid plate armor had its weaknesses. A full suit of armor covered the warrior from head to toe. It protected against most cuts and blows from swords and spears. But a well-aimed bolt from a powerful crossbow could pierce it.

Gaps were another problem. When a knight stretched an arm or leg, the joints where armor plates overlapped would pull apart. This would open up gaps between the plates, which opponents would look for. They would try to drive a sword through the gap and into the chest or arm or leg beneath. That's why knights wore chain mail beneath their plate armor. Sometimes the mail protected these deadly gaps. Sometimes it didn't.

Super-Tough Skin

In the 2000s, military body armor is far superior to the metal armor knights wore. Military body armor is made with materials stronger, lighter, and more flexible than iron. The fabric known as Kevlar, for instance, was first developed to make car tires stronger. Later on, scientists found they could turn it into body armor.

These U.S. Army soldiers are wearing Kevlar vests to protect them from bullets.

Kevlar is soft body armor. It works like an extra, super-tough layer of skin. Kevlar does not send bullets bouncing away. Instead, this soft skin acts like a net. It catches the bullet and breaks its power into little bits. These bits of energy scatter through the armor in all directions. That way, the bullet loses its power to seriously hurt the soldier.

Saved by Dragon Skin

One day in June 2005, a U.S. military worker in Iraq had on a vest of soft body armor. It was made of material called Dragon Skin. The vest covered his chest and the upper half of his legs. During the day, he was fired on. At the end of the day, he took off the vest and saw damage to it.

"I did not even KNOW that I was hit twice until I took off my tactical vest—this was about two hours after the contact—and saw . . . that I had taken hits in the back. It was only then that we took a close look at my body armor that we realized I was hit twice by an AK-47."

The military worker was not injured.

Of course, this armor will not protect against everything. Soldiers still are hurt and killed in combat. But modern body armor has saved many soldiers from life-threatening injury.

A Death Ray

For every deadly weapon in war, the military works to create a technology to defend against it. Sometimes that defensive technology is a new version of the deadly weapon itself. Antimissiles are missiles made to seek out and shoot down missiles the enemy has launched toward you.

The U.S. Air Force is testing a new antimissile system called the Airborne Laser (ABL). The ABL is a directed energy weapon, like the ADS heat ray. It will be operated from a Boeing 747 aircraft.

The aircraft will fly continuously over a known enemy missile site in a figure-eight pattern. The aircraft will have heat-seeking infrared sensors to detect an enemy missile the moment it is launched. When a missile is launched, the information will be fed into the plane's computer. The computer quickly will direct ABL's three lasers to target and fire on the rising missile.

The first two lasers identify the missile, track it, and send the information back to the computer. Then the third one, a chemical laser, shoots down the target. A chemical laser gathers its energy from chemical reactions and sends this chemical energy out in a beam. The ABL laser is classified as a death ray, since it has the power to destroy its target.

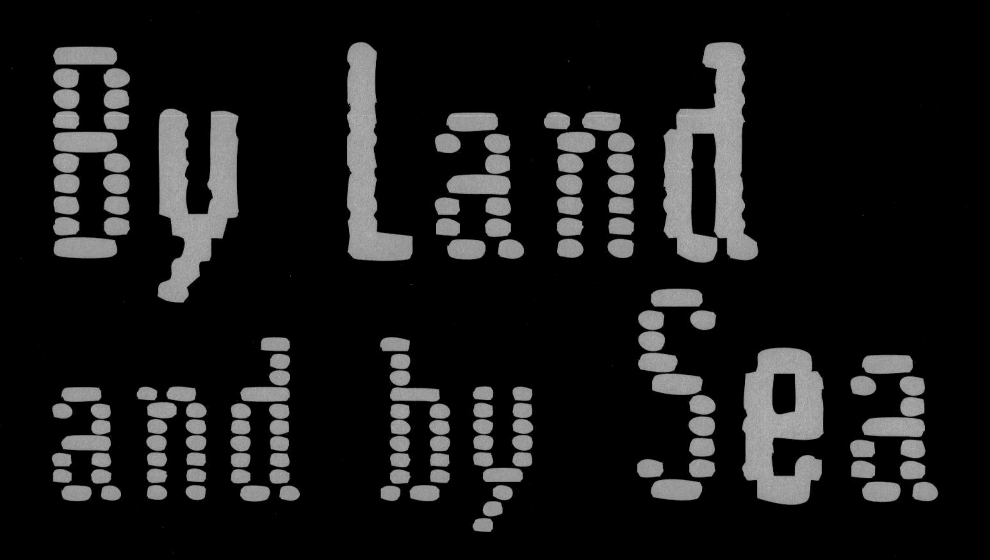

By Land and by Sea

Sometimes soldiers on land fight from moving vehicles. Four thousand years ago, soldiers would go into battle on chariots, two-wheeled wooden carts pulled by horses. Two or three warriors would stand in the cart holding spears, bows and arrows, shields, and the horses' reins.

This engraving from the 1600s shows an ancient chariot, from about 1300 B.C., with two soldiers in armor riding in it.

Armored Fighting Vehicles

Modern-day soldiers use armored fighting vehicles (AFV) to fight on land. An AFV is armed with weapons and protected by armor. Most AFVs are equipped for driving across hilly, rough land.

Some AFVs are tanks, which run on steel tracks. Battle tanks come equipped with powerful guns for firing directly at the enemy. In the 2000s, the M1 Abrams is the main U.S. battle tank. The tank's main gun is used to pierce the armor of enemy tanks. It can also be pointed upward to fire at aircraft. The Abrams is also equipped with machine guns and smoke grenade launchers. The grenades produce a thick smoke to block the enemy's vision in battle.

To protect the soldiers inside, the Abrams is fitted with armor made of several layers of steel and ceramics, or hardened

IT'S A FACT

The Abrams tank comes with GPS, laser, and infrared systems for locating and locking in on targets.

clay. For added protection, some Abrams tanks also are fitted with panels of explosive reactive armor (ERA). High explosives are sandwiched between metal plates on the tank's sides. When hit by enemy shells, these plates blow up. That sounds as if the ERA would hurt, not help, the passengers. But explosive reactive armor blows outward. This stops the enemy shell from piercing the tank and harming the warriors inside.

Armored Personnel Carriers

Armored personnel carriers (APCs) are armored fighting vehicles used to safely transport infantry soldiers on the battlefield. Some are tanks, with tracks. Others are trucks that run on wheels.

Most APCs carry from six to ten passengers. Most are also amphibious—they can travel on land or in water. In water, tracked APCs are propelled along by the movement of their tracks. Wheeled APCs move along by propellers or jets of water.

Most APCs can move on land or in water.

The U.S. military's newest APC is a line of eight-wheeled vehicles known as the Stryker family. Strykers are equipped with new features to make them safer and easier to run. In case of fire inside, a fire extinguishing system automatically comes on. The vehicles have a special tire inflation system. The driver can quickly change tire pressure to fit the type of surface the Stryker is driving on, from smooth asphalt to rough rock or thick mud. Like submarines, Strykers also come equipped with periscopes. These instruments allow the driver and commander to see the landscape while remaining safely inside the armored vehicle.

Military Technology at Sea

Ancient sea battles were fought in ships called war galleys. Many galleys had a mast with a sail. But most of the power came from muscles.

The Strykers

The Stryker armored vehicles are named in honor of two heroic U.S. soldiers killed in action. Private First Class Stuart S. Stryker received the Medal of Honor for his actions during World War II. Specialist Fourth Class Robert F. Stryker (unrelated to Stuart) received the same award for his actions during the Vietnam War (1957–1975).

Stuart S. Stryker's award citation tells how he led the men under his command "in a desperate charge" through a "hail of bullets." Stryker himself was killed, but his "gallant . . . action in the face of overwhelming firepower" led his company to capture more than 200 enemy soldiers and free three American prisoners.

Robert F. Stryker's citation states that during a fierce battle he "observed several wounded members of his squad in the killing zone of an enemy claymore mine. With complete disregard for his safety, he threw himself upon the mine as it was detonated."

U.S. soldiers patrol the streets of Mosul, Iraq, in Stryker vehicles in 2006.

Men rowed with oars to propel these long, low ships through the water.

This painting from the 1800s shows a war galley from 800 B.C. The ship was powered by men rowing and by a sail.

Modern military ships are powered by engines. Some use nuclear power, the same type of energy unleashed by nuclear bombs. Nuclear-powered engines can travel very long distances without refueling. The largest, most powerful nuclear-powered ships are aircraft carriers. These huge vessels are like military air bases at sea. They have a long, flat deck like an airport runway. The largest carriers carry dozens of aircraft and thousands of crew members.

Some nuclear-powered craft travel underwater. The USS *Nautilus* was the world's first nuclear-powered submarine. In 1958 it became the first ship to travel underneath the ice cap at the North Pole, a distance of more than 1,800 miles (2,897 km).

The USS *Nautilus* enters New York Harbor during a routine training operation in 1956.

To accomplish this feat, the *Nautilus* had to cruise deep underwater for four straight days. Subs cannot carry large supplies of oxygen. Every bit of space on a submarine is precious. Storing all that oxygen would take up too much room. Then how did the *Nautilus*'s crew members get enough oxygen to breathe?

A molecule of water (H_2O) is made of atoms of hydrogen (H) and oxygen (O). Nuclear submarines use electricity to split apart water molecules into hydrogen and oxygen gases. They send the oxygen into the air of the ship for crew members to breathe. The hydrogen gets thrown away.

The First Submarine Mission

On September 7, 1776, a historic military event took place. The first submarine tried to sink an enemy ship. *Turtle* was powered by a hand-cranked propeller. The one-man sub was simple, but it used some of the same scientific principles that modern submarines use. To send *Turtle* down, water was pumped into the hull by a hand pump. The water was pumped out to bring it back up.

Turtle's inventor, David Bushnell, had a daring mission in mind. During the American Revolution, the American colonists fought against the British. The British controlled New York Harbor with warships. What if *Turtle* could secretly attach a bomb to one of them?

The American soldier chosen for the mission, Sergeant Ezra Lee, waited for dark. Then he climbed into *Turtle* and set off under the surface of New York Harbor.

Lee cranked his way out to the British warship *Eagle*. The bomb was a keg of powder with a time fuse. Lee tried and tried, but he could not cut through *Eagle*'s copper-plated hull to attach the bomb. The mission failed. But Lee, Bushnell, and *Turtle* became part of military history.

This drawing shows the new Virginia-class attack submarine being built for the U.S. Navy.

Modern Nuclear Subs

Modern nuclear submarines are smaller, faster, and better armed than the *Nautilus*. The Virginia-class nuclear submarines are good examples. The U.S. Navy plans to eventually build 30 of them. The first one, the *Virginia*, was launched in 2003.

How can these subs be used in warfare? The *Virginia* is equipped with tubes for launching cruise missiles and torpedoes. It also has a special chamber in its hull to hold a minisubmarine. The minisub can carry teams of warriors—special warfare forces from the U.S. Navy or U.S. Marines. It can be launched underwater right from the chamber. Then the mini-sub can deliver the special forces teams to shore wherever they are needed to fight.

In the past, submarines used periscopes to get a view above the surface while remaining hidden underwater. Periscopes use a series of mirrors and prisms to reflect the view down a tube to the inside of the sub. The Virginia-class nuclear subs use photonic masts instead.

A photonic mast contains scientifically advanced viewers and sensors. A low-light TV system brings down a view of the surface. Infrared and laser sensors locate and home in on targets.

Up in the Air

Benjamin Franklin (1706–1790) was famous for his revolutionary ideas. He thought hot-air balloons should carry bombs to drop on enemies during wars.

Franklin's idea never got off the ground. But in the 1900s, the Wright Brothers invented a motor-powered airplane. Not long after that, warplanes were used as bombers—they dropped bombs on their targets.

British and German fighter planes had aerial dogfights during World War I.

Fighter Planes

By World War II, bombers were used to destroy enemy cities and bases. But bombing missions were very dangerous. Bombers were large and slow. They were easy prey for smaller, quicker fighter planes.

Fighter planes provide protection. Modern fighters are equipped with high-powered guns. They also have missiles for air-to-air combat—shooting down enemy planes. Fighter planes can also provide close air support. They support infantry by striking enemy targets on the ground.

Modern fighters can travel at supersonic speeds. That means they travel faster than the speed of sound. Sound travels at 761 miles per hour (1,225 km/h), or Mach 1. The U.S. Navy's top fighter, the F/A-18 Super Hornet can reach a speed of Mach 1.8—almost twice the speed of sound.

The Barrel Roll

Tactics are important tools of war. Every fighter pilot knows a tactic called the barrel roll. It's a defensive maneuver to avoid an enemy plane that is coming at you from the rear. (Aircraft are most at risk of being shot down from behind.)

Pilots try the barrel roll when their attacker is "locked in" on their tail. The pilot will make a quick break in another direction. But only for a split second. Then the pilot rolls the plane in the opposite direction. The move changes direction and slows the airplane down. If the barrel roll catches the attacker by surprise, he won't have time to match it. Instead, he will fly right past. If it all works correctly, the two planes will have switched positions. Suddenly, the hunted becomes the hunter.

Invisible Planes

If you were a fighter pilot, wouldn't you love to fly a plane that the enemy couldn't see? Then they couldn't attack you. The U.S. Air Force has a supersonic fighter plane that is nearly invisible. It's the F-22 Raptor.

The Raptor is nearly invisible because it has something called stealth. How does stealth work?

The key is design. In a war, each side keeps scanning the sky, looking for enemy planes. A plane equipped with stealth is designed to fool enemy scanners. It is nearly impossible to see, feel, or hear.

Ordinary planes look like black dots in the sky. But not the Raptor. A special gray paint makes it blend in with sky and clouds when seen from below. When seen from above, it seems to disappear into the landscape below. That's how it fools eyes.

Stealth also can fool ears. The Raptor and other stealth planes are quieter than normal jet-powered aircraft. Special systems reduce the sounds their engines make. It's nearly impossible to hear.

This F-22 Raptor uses stealth technology to be invisible in the air.

Fooling Radar

The most important scanner to fool is radar. Radar scanners send out bursts of radio waves. When they hit something solid, like a plane, the waves bounce back to the scanner. Computers use this bounce-back information to find the plane. Then antiaircraft weapons can shoot it down.

But F-22 Raptors hide from scanners because they are made of special metals and curved surfaces. This special design scatters or absorbs radar waves.

Another kind of scanner detects the heat from an object. It detects the heat from the exhaust that a plane's engines send out. Aircraft engineers have found ways to cool down engine exhaust gases. That way heat scanners won't know that the airplane is up there. The F-22 Raptor is the ultimate in military aircraft technology.

A Chameleon Plane

A chameleon is a lizard that can change its color to match its background. Imagine a plane that could do what chameleons do. Flying over a forest, its upper surface would turn green to match the trees below. On a cloudy day, its lower surface would change to match the clouds above. No matter what the land or sky looked like, this plane would always blend in.

Sound like science fiction? For now it is.

But scientists are working with special paints that can change color electronically. Computers and sensors would automatically change the color to match the land or sky. And that would make the plane superstealthy: invisible from any angle.

The U.S. military is using military technology to make future wars safer for soldiers. What is the safest way for a soldier to fight a war? That's easy: to be absent from the battlefield altogether.

But then who would do the actual fighting? Robots would. And who would control them? Humans with computers would control them from a safe distance. This may sound like a video game, but it's war: a new kind of war.

PackBot

Some of these robots are just plans that have not been carried out. Many are still in the experimental stage. Scientists are still testing them.

Others are already in use. One is the PackBot Scout, the world's first battlefield robot. PackBot weighs about 40 pounds (18 kg). Human soldiers can carry it in a backpack and assemble it in a few minutes. PackBot looks like a minitank with a mechanical arm.

The soldiers control the PackBot Scout with laptop computers, using PackBot's GPS receiver to guide it. The soldiers can send the robot almost anywhere to look for danger. It can search buildings in cities. In the countryside, it can travel up steep hills and deep into dark caves.

PackBot's mission is to look ahead and search for information. As PackBot travels, soldiers watch on computer screens. PackBot's video cameras give soldiers a safe look at what lies ahead. And PackBot's arm can pick up and examine objects that look suspicious.

FACT

PackBot Scout fought with U.S. ground troops in 2002 in Afghanistan and later on in the Iraq war.

(Above) A technician controls a PackBot robot via remote control in Iraq. *(Below)* A PackBot at work during a demonstration

Scouting for Danger

The PackBot Scout looks for dangers of all kinds. In Iraq the greatest danger to troops are improvised explosive devices (IEDs). IEDs are homemade versions of EFPs. They are used by the insurgents.

Most IEDs are exploded by remote control. Insurgents wait until soldiers or their vehicles are near the hidden bomb. Then insurgents press a button on an electrical device, such as a cell phone, to explode it.

IEDs can be hidden anywhere: beneath a rock or inside a soda can, a paper bag, a truck, or a car. Some have even been hidden inside dead animals. IEDs are not very sophisticated, but they are also effective. Many U.S. soldiers killed and wounded in Iraq have been victims of IEDs.

Many more soldiers would have been killed and wounded if not for the PackBot Scout. The Scout can use its claw to disarm or explode IEDs. Sometimes the robot is damaged or blown up. But machines are replaceable. Human lives are not.

Robots in Iraq inspect a box that could be a bomb in a street in Baghdad (*top*) and to check out the site of a roadside bomb in Basra (*bottom*).

Robots with Guns

What about armed robots? The U.S. military is testing a robot that carries a remote-control machine gun. It's called the Talon. The Talon is 2.5 feet (0.8 m) tall and moves on treads like a tank.

Then there's the Gladiator, a bigger armed robot also in the testing stage. The Gladiator is 4 feet (1.2 m) tall and weighs 1,600 pounds (726 kg). It can fire several different weapons. Like the Talon, it also moves like a tank.

U.S. troops are using the Talon robot in Iraq.

Talon and Gladiator are remote-controlled robots that work on the ground. What about robots that work in the air?

Flying Robots

Robot warplanes are called drones. The U.S. military has been using these remote-controlled aircraft since the Persian Gulf War (1991). Like ground-based robots, drones are controlled by soldiers with computers.

Most drones are small. Only two or three people could fit in one. That is, if it carried human passengers. Drones have no windows because there is no one inside to look out. Instead, drones have onboard video cameras. Soldiers working in computer stations on the ground view the ground through the drone's cameras. Then soldiers can warn those in the field of any dangers ahead.

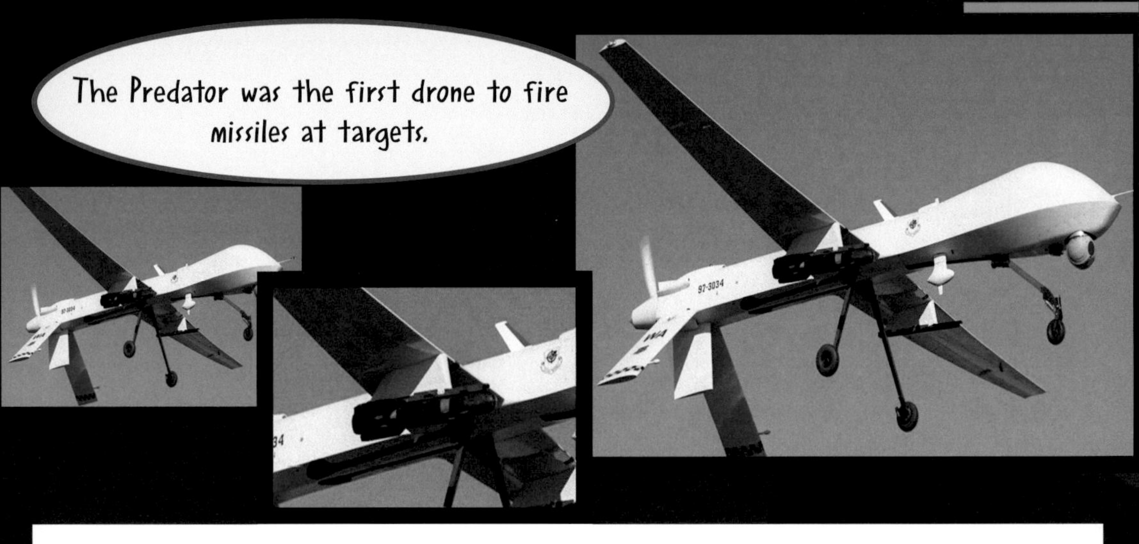

Drones sometimes fly with weapons as well. In the Afghanistan war, U.S. drones called Predators fire Hellfire missiles on targets.

The U.S. military is making more and more drones. Experts say that within 20 years, nearly half the U.S. air attack missions will be flown by robot aircraft.

The Future

More robots are on the way. One is an ambulance that drives itself to pick up wounded soldiers on the battlefield. Another is a robot with a mechanical arm that can drag a wounded human soldier to safety. There's even a four-legged robot in the testing stage that can carry 88 pounds (40 kg) of equipment across any kind of battlefield. And there's more to come. One day, perhaps, human-controlled robots will do all the battlefield jobs that soldiers must do now.

From muscle power to computer power, military technology continues to change the way armies fight wars. Wars continue to be fought. It is estimated that nearly 200 million people were killed in armed conflicts during the last century. More than 700,000 people were killed in 2002 alone.

Minidrones

Imagine an aircraft small and rugged enough to carry around in a backpack. Meet Dragon Eye. This minirobot plane is made of lightweight fiberglass and Kevlar. It weighs about 5 pounds (2 kg). From wingtip to wingtip, it's just 3 feet (1 m) wide.

Dragon Eye comes in five pieces: a body, a nose, a tail, and two wings. Two soldiers can put it together and launch it by hand in about 10 minutes.

The soldier who controls Dragon Eye uses a wearable computer, a computer attached to the soldier's uniform. A screen on the arm or vest shows a map of the area nearby. By clicking on the map, the soldier tells Dragon Eye how high to fly and where. Dragon Eye can stay in the air for about 45 minutes at a time. With a few more clicks, the soldier can tell Dragon Eye when to return.

Dragon Eye can fly at 35 miles per hour (56 km/h) at an altitude of 1,000 feet (305 m). Video cameras in its nose send back images of the ground. The soldiers below see what Dragon Eye sees. They watch the view on their computer screens.

Dragon Eye is a stealth drone. It is so small that radar scanners see it as a bird. So enemy soldiers don't even know this amazing little plane is up there, looking down on them for U.S. troops to see.

All through history, conflict and violence have been part of human life. We can only hope that one day people will find the patience and wisdom to put down their weapons and settle their differences peacefully.

aircraft carrier: a large warship that carries planes and has a long flat deck for planes to takeoff and land

air-to-air: combat between two or more aircraft

archaeologist: a person who digs up ancient things, such as bones and weapons, to learn about ancient cultures

armor: tough, protective clothing used by soldiers in battle. Also, protective plates used on vehicles such as tanks, ships, and aircraft

bomber: an aircraft that drops bombs on enemy targets

bullet cartridge: a metal container holding a bullet and gunpowder, fired by a gun

chain mail: flexible armor made of metal rings

chariot: a two-wheeled vehicle carrying soldiers, pulled by horses

crossbow: a bow mounted crosswise on a stock, which shoots arrows called bolts

drone: an unmanned aircraft that is operated by remote control

fighter plane: a fast, maneuverable aircraft that can fight other aircraft, as well as attack targets on the ground

galley: a long warship that is powered by oars

global positioning system (GPS): a system of satellites and receiving devices used to locate positions on Earth

gunpowder: a mixture of potassium nitrate, charcoal, and sulfur used to create explosions

improvised explosive device (IED): a homemade explosive weapon

infantry: soldiers who fight on foot

infrared: light that is not naturally visible to the human eye but can be made visible by technology

Kevlar: a high-strength, bulletproof fiber used to make soft body armor

military technology: the tools that fighting forces use to attack and defend

precision-guided bomb: a bomb that can be guided to its target by remote control

radar: a device that uses radio waves to locate distant objects

stealth: technology used to make aircraft invisible to radar and other detection methods

supersonic: faster than the speed of sound (761 miles per hour/1,225 km/h)

warhead: an explosive device, usually attached to a bomb or a missile

wearable computer: a small portable computer that is worn on the body as it is being used

Selected Bibliography

Alvarez, Cristobal. "A Summarized History of the Development of Military Technology." *The War Scholar*. 2000–2003. http://www.warscholar.com/Year/TechnologyOutline.html (March 18, 2007).

CNN. "A Day of Sirens, Bombs, Smoke and Fire." *CNN/com/World*. March 31, 2003. http://www.cnn.com/2003/WORLD/meast/03/21/sprj.irq.aday/ (March 18, 2007).

———. "Special Report: War in Iraq: Forces, Weapons." *CNN.com.* 2005. http://www.cnn.com/SPECIALS/2003/iraq/forces/weapons/index.html (March 18, 2007).

Cotterell, Arthur. *Chariot: From Chariot to Tank, the Astounding Rise and Rall of the World's First War Machine*. New York: Overlook Press, 2005.

Cowley, Robert. *The Reader's Companion to Military History*. Boston: Houghton Mifflin Company, 1996.

Crane, David. "Under Fire: Field Report on Pinnacle Body Armor." *Military.Com*. 2005. http://www.military.com/soldiertech/0,14632,Soldiertech_PArmor2,00.html (March 18, 2007).

Davis, Paul K. *100 Decisive Battles from Ancient Times to the Present*. New York: Oxford University Press, 2001.

Golan Group. "The Corner Shot Weapon System." *Golan Group*. 2007. http://www.golangroup.com/products-cornershot.shtml (March 18, 2007).

Hambling, David. "Instant Expert: Weapons Technology." *NewScientist.com news service*. September 4, 2006. http://www.newscientisttech.com/channel/tech/weapons/dn9980 (March 18, 2007).

Murray, Stuart. *Atlas of American Military History*. New York: Facts on File, 2005.

Vizard, Frank. *21st Century Soldier*. New York: Time, 2002.

Further Reading and Websites

Ancient Greek Warfare for Kids

http://www.historyforkids.org/learn/greeks/war/index.htm. This site includes lots of information about ancient civilizations as well as ancient warfare. Check out the section titled "How Greek Men Fought Wars (Armor, Weapons, Tactics)."

BBC: Children of World War 2

http://www.bbc.co.uk/history/ww2children/. Find out what life was like for children caught in the middle of a world war. This site is all about children in Britain during World War II. It includes letters from children of the time.

Bohannon, Lisa Frederiksen. *The American Revolution*. Minneapolis: Twenty-First Century Books, 2004. Read about the eight years of bloody war that ended in independence for the United States of America.

Dartford, Mark. *Bombers*. Minneapolis: Lerner Publications Company, 2004. Discover more about bombers, aircraft that deliver bombs, missiles, and rockets with the help of high-tech targeting equipment. They can take out the enemy's planes, ships, or other military targets with precision from a long way away.

____. *Fighter Planes*. Minneapolis: Lerner Publications Company, 2004. Modern fighter planes cruise at more than twice the speed of sound. Their computer equipment can spot and identify enemy planes from miles away. They are equipped with missiles that can lock on to an object, track it, and put it out of action. Discover more about these amazing machines.

____. *Missiles and Rockets*. Minneapolis: Lerner Publications Company, 2004. Learn about missiles and rockets, the ultimate attack weapons. Equipped with a hi-tech targeting system, a modern missile can travel thousands of miles and still hit its mark within inches. Missiles and rockets are a nation's strongest first-strike option.

Doyle, Kevin. *Aircraft Carriers*. Minneapolis: Lerner Publications Company, 2004. Aircraft carriers are floating air bases, ready to deliver military power where it is needed most. Discover more about the most powerful ships on the high seas.

____. *Submarines*. Minneapolis: Lerner Publications Company, 2004. Powered by nuclear engines, submarines can sail for years without refueling. Armed with missiles and torpedoes, submarines are a navy's toughest offensive weapon. Discover more about these underwater weapons, from the deadly German U-boats of World War I to modern high-tech hunter-killer subs.

Future Weapons
http://dsc.discovery.com/fansites/future-weapons/weapons/zone1/weapon-zone-1.html. This site is part of the Discovery Channel, dealing with the latest in military technology.

Goldstein, Margaret J. *World War II: Europe*. Minneapolis: Twenty-First Century Books, 2004. Fought in the deserts of North Africa, the mountains and plains of Europe, and the waters of the Atlantic and the Mediterranean, World War II was the biggest war the world had ever seen—and the biggest victory the United States had ever won.

Military Technology News
http://www.spacewar.com/Military_Technology.html. This site features the latest news in military technology, updated daily.

Sherman, Josepha. *The Cold War*. Minneapolis: Twenty-First Century Books, 2004. This book describes the Cold War, which was led by the superpowers the United States and the Soviet Union and marked by the building of the Berlin Wall, the Cuban Missile Crisis, the space race, and the development of nuclear weapons.

Time for Kids: America at War
http://www.timeforkids.com/TFK/specials/iraq/0,8805,424876,00.html. This site is about the Iraq war. It includes a timeline of events, information on the nation of Iraq and its neighbors, and biographical sketches on the leaders of the conflict. It also includes articles by kid reporters and letters from kids about the war.

Woods, Michael, and Mary B. Woods. *Ancient Warfare: From Clubs to Catapults*. Minneapolis: Twenty-First Century Books, 2000. This book explains how various ancient weapons were invented and how they were later changed to be more portable and powerful.

Index

Photo Acknowledgments

The images in this book are used with permission of: © Todd Strand/Independent Picture Service, background image on p. 1 and even pages; © Johancharles Van Boers/AFP/Getty Images, p. 5; © Bettmann/ CORBIS, pp. 6, 14; National Archives, pp. 7 (W&C 163), 19 (W&C 1242); Library of Congress (LC-USZC4-1766), p. 8; © Arthur Turner/Alamy, p. 10; © Uriel Sinai/Getty Images, p. 11 (top); © INTERFOTO Pressebildagentur/Alamy, p. 11 (bottom); The Department of Defense, p. 15; TSGT. Bob Simons/The Department of Defense, p. 17; Christopher Senenko/U.S. Navy Photo, p. 18; © Time & Life Pictures/Getty Images, pp. 20, 21; U.S. Army, p. 23; © Hulton Archive/Stringer/Getty Images, pp. 25, 29 (bottom), 32; Eric Steen/The Department of Defense, p. 26; Phan Paul Polach/The Department of Defense, p. 27; John M. Foster/U.S. Army, p. 28; The Art Archive/Bibliothèque des Arts Décoratifs Paris/Dagli Orti, p. 29 (top); U.S. Navy Photo, p. 31; Tech. Sgt. Ben Bloker/U.S. Air Force, p. 34; Spc. Jonathan Montgomery/U.S. Army, pp. 37 (top), 39; AP Photo/U.S. Army Handout, p. 37 (bottom); © Ahmad Abdel Razak/AFP/Getty Images, p. 38 (top and inset); © Essam Al-Sudani/AFP/Getty Images, p. 38 (bottom); U.S. Air Force, p. 40.

Front cover: Chief Journalist Dave Fliesen/U.S. Navy (top); Staff Sgt. Aaron Allmon II/U.S. Army (left); Spc. Jonathan Montgomery/U.S. Army (right); © Todd Strand/Independent Picture Service (background). Back cover: © Todd Strand/Independent Picture Service.

About the Author

Ron Fridell has written for radio, television, and newspapers. He has also written books about the Human Genome Project, including *Decoding Life: Unraveling the Mysteries of the Genome*; about genetics, including *Genetic Engineering* in the Cool Science series; and the use of DNA to solve crimes. In addition to writing books, Fridell regularly visits libraries and schools to conduct workshops on nonfiction writing.